EXTRATERRESTRIALS

What & Who They Are.

Dr. Friday E. Abu

*This book is dedicated to all those who want
to know more about the existence of other
intelligent lives out there in the universe.*

CONTENTS

PREFACE

Extraterrestrials have visited the entire world, they are still visiting, they are among us and they will keep wielding the impact they want on the world. Around the world, Unidentified Flying Objects (UFOs) have been sighted and they will keep being sighted. The Unidentified Aerial Phenomena (UAP) will not cease; it will continue.

The Cosmos has to send teachers, advanced technologists, healers and others from time-to-time. The UFOs are being seen but, how about the beings dropped off by these advanced crafts to mingle with the human population? Who are these beings? Some were seen; others were noticed by their presence which was just felt. However, the advanced technological crafts far beyond humanity's present and near-future invention capacity are being seen across the globe. Are they here to invade the earth? Wait and let us find out. What concern has lions with ants? These guys are far advanced.

All over the world, the UFOs were not just seen;

some entities admittedly told humanity – directly or indirectly – that they came from the stars via these crafts. This truth has come to be accepted because the ancient man, in notably old writings e.g. the Mahabharata mentioned crafts used by the supposed 'gods' that visited them from the sky. Supportive to this claim in this modern time and era, is the nascent discovery of some habitable exoplanets around a good number of stars. If humanity and the earth are just too young on the evolutionary scale for man to traverse the distances between the stars, then, another civilization twice or possible thrice the age of humanity may not be young and therefore, should well have attained the technical know-how to beat the gaps separating the stars. If not, they would not have been seen on and/or around the earth.

INTRODUCTION

When one of the greatest presidents in the history of the United States, President Franklin D. Roosevelt, wrote a memoir to the special committee on non-terrestrial science and technology in February, 1944, he made it clear that non-terrestrial know how in atomic energy must be used in perfecting super weapons of war… [I'll stop there but, his statement did not stop there. Let me prevent any digression and ambiguities. Thanks goodness! The world is now enjoying peace.] He landed openly, the entire world on a reality, 'the reality that our planet is not the only one harboring intelligent life in the universe,' as put by that memoir. Here, he was not babbling, neither was he blabbing. Of course there are intelligent lives out there and some of these lives have gone further than humanity in technological feats that man has been learning from. Let us face the facts; the beings with the technical know-how to visit earth are above humanity in every facet and aspect of scientific advancement. That is why even the then president of the US, Roosevelt, deems it fit that we should learn from them.

Right from antiquity, since a very long time ago, the earth has been known to have harbored beings that have higher capacities than the conventional earthly beings. Some of these beings are human-like (humanoid), others possessed human-animal features combined, few had human-and-unknown features on their bodies and the rest had complete animal features. Among those with complete animal features, there are some with human voices projecting out of them; others had extremely uncanny capabilities when compared with mere animals with which they shared visible features. Yes, there were abilities known to be 'Extra' about these beings. These beings were recorded to have recognizable and proven various capabilities that the ordinary humans or creatures did not possess. Indeed, they are the Extraterrestrials (ETs, for short).

Some of the ETs told humans that they are not of this planet. When they were asked where they came from, they simply pointed to the stars. Truly, they did came from the stars; a distance far beyond our present technological breakthrough. Surely, when talking about their technology which has been able to transport them from another star system to ours, there's equally something 'Extra.'

CHAPTER I
Their Coming to the Earth

The coming to the planet earth by the Extraterrestrials has fascinated men and women from antiquity till this age. This is because the traversed distances to reach the earth on a linear scale are definitely lengthy, ordinarily. However, on a cosmic scale, the closest star, Proxima Centauri, is actually located at our star system's backyard. That is to say it is a short distance on a cosmic scale. But, using our current technology to reach that closest star to us will take humanity about 6,000 years, if not more. Obviously, humanity has a lot of technological breakthroughs to make. Why? This is because these beings (the ETs) that visited earth and are still visiting, have truly gotten exceptionally further ahead technologically and that is the reason for their capacity to boldly embark on journeys that would cost man about 6,000 years, going by present humanity's best manned space vehicle. I want to strongly believe that these beings would not spend much time in their journeying around the Cosmos; after all, the technological feats displayed by their crafts (the UFOs) are unbelievably stupendous. Their crafts:

- Have no exhaust trail.
- Can split into 2 or more.
- Become invisible while in motion.
- Attains almost speed of light acceleration from any motionless position.
- Can be covered by light hue all over.
- Move from water into air and vice versa.

Humanity has not attained any of the above characteristics of the crafts of these beings mentioned in our earthly crafts. Of course, President F.D. Roosevelt was right. They are ahead of us technologically and that is why he asked us to acquire their technical know-how. It is the same reason behind their being able to visit the earth easily.

Humanity has just started talking about warping the space time fabric; a technological feat that has not gotten off the ground yet. ETs seemingly perfected this technology and are using it to cover even distances that are lengthy on the cosmic scale. There are star systems out there that are older than our sun. So, I expect that if such stars are harboring intelligent life, that life should exceed man in technological advancements. If man is about developing a working warp drive, could it be that another civilization that existed before man and on another planet perfected this technology a long time ago? Yes, it could!

Some beings, from ancient fables, pointed at the Sirius, Orion and Pleiades star systems as the places they came from. They must have discovered that their technology could make it to the earth and they came. If not, they would not have tried it. In most instances, their crafts have been seen defying the earthly laws of Physics. I have heard, times without number, that a visibly glaring UFO in the sky just vanished in a terribly high speed.

[Please, note that before the comparison to be made below, UFOs refer to crafts belonging to extraterrestrial origin. I will compare two different things that people see in the sky quickly below.]

For the sake of comparing and appreciating the level of technological advancements between man and these higher beings, I classified these crafts apparently seen and reported by people into 2 groups:

1. Unidentified Physics Congruent Objects (UPCOs)
2. Unidentified Physics Defying Objects (UPDOs)

The first set, UPCOs agree with and operate in accordance with our earthly Physics. Once in a while, they are seeing above cities, have exhaust trails and do not defy the laws of Physics in any way. Some individuals like me do strongly believe they are earthly mechanical works trying to mimic non-terrestrial inventions. They are fast but, not an unearthly speed.

The second set above, UPDOs disagree completely with every known law of Physics, treats gravity as though it does not exist, with a speed that is almost that of light and these are the true non-earthly inventions that are inexplicably mysterious to every observing eye. Could it be that the inventors of these crafts have outdone the laws of Physics? The UPDOs that do not leave any exhaust trail may be harnessing the cosmic energy to power the crafts. Better yet, they may be powered by dark matter of the universe. These forms of alternative sources of energy which can replace carbonated fuel, are just being talked about and not yet experimented on by the earth. What if these advanced civilizations that existed before humanity mastered these sources before now, too? There's every possibility!

The great Physicist, Albert Einstein, mentioned the existence of worm holes and the earth is still searching for them, scientifically. There are scientific speculations already pointing at the possible reality of these cosmic short cuts to other universe and star systems already. Since humanity is young, I want to believe that these higher beings, older on the evolutionary scale than us, must have glaringly pin-pointed the right places having worm holes in the universe. Via these tunneled cosmic short cuts, they are easily reaching us. So, it is not only by the speed of crafts but, also the use of galactic holes, are we being easily reached by the ETs.

The cars have their roads, the trains have their rails, the ships have their courses and the planes have their routes. Has humanity been able to map out the galactic highways to other star systems? What if these highways are there but, we are not seeing them because of our nascent technology? May be these beings are making use of a galactically and strategically mapped-out highways for spaceships. It could be that taking such a mapped-out galactic trajectory, other planets and star systems could be reached i.e. if such truly exists but, yet to be known unto man.

CHAPTER II
Who Are They & How Do They Look?

The Extraterrestrials that have made it to surface of the earth at various times were very brilliant, intelligent and wise beings who baffled the men and women of their time with unmatched capacities at different levels. The humans they visited were awestruck because of unpredictable prowess in mental displays from these ETs. Some of them taught medicine, technology and other aspects of sciences. Some taught architecture, building and even supervised the structural completion of great temples and buildings which man would not have built, if they were not involved. Due to this mentality and brain usage, having existed before man, ETs have a greater level of brain and mind evolution. Their skulls have adaptively grown bigger whenever they are looked at – especially in those that are upright like man – that one may think that their heads are disproportionately bigger when compared to other body parts.

The body of these beings may be mistaken to be naked but, they are actually clad in clothing that seeming tightly fits their body. Their technological advancements enabled the development of highly stretchy and glossy suits that wrap their bodies almost in entirety. This materials look like leather with translucent external finishing.

When observed, within few seconds, it looks as though the muscles on some ET races are depleting or already in depleted state. It is an evolutionary stage arrived at due to lack of muscle engaging activities brought about by technological advancements and use of machines for muscle-tasking jobs on their home planets. These ETs developed various machines to handle arduous muscular tasks and over time, evolutionary trend began presenting them as lanky non-muscular figures. Man may evolve to this stage one day, if we deploy robots to do all things while we lie down to sleep.

ETs walk and move like man, for those among them who walk on two feet. However, speed of movement differs depending on the characteristics of the feet. For instance, the VaDoma tribe of Zimbabwe with a feet condition called ectrodactyly (middle toes absent and the two outer ones which are the present, are turned in) said their first ancestors came from the star system of Sirius.

These ET ancestors of theirs had feet looking like those of Ostriches, according to a reputable source. These Sirius originated ETs could walk on both

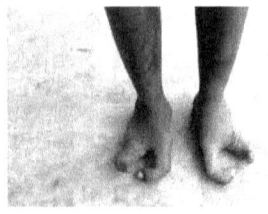

legs and they were equally exceptionally good tree climbers. The Sirius ETs intermingled with women in this region and birthed children. That is how humans from this tribe started having the condition. Indeed the features man share with ETs cannot be completely similar, as seen from this tribe's example.

Materializing and dematerializing (what has come to be known as 'Transdimensional' existence) is displayed by ETs sometimes. This capacity shows a greater deal of evolution than man. To be able to fade-out and seemingly vanish out of the sight of man and still be there, is a level that man will not attain very soon. But, these beings have attained it. It is actually being able to operate in a 3-dimensional world and then, transcend to a higher dimensional reality consciously, while standing on the earth. They do this mostly when they feel threatened or when they feel a man is frightened by their presence.

ETs can sense our moods. These beings sense our: anger, fear, pain, depression etc. and they sense them quickly. Depending on the mood sensed, they either continue their interaction or disengage. If they sense any mood not palatable to their interaction, they quickly vanish, most of the times.

It is no news anymore that ETs can talk.

Most of the time, they talk telepathically. When human abductees were disbelieved by the scientific community, they were subjected to regressive hypnosis. It was discovered that their accounts were correct. During the hypnosis sessions, these abductees were asked how these beings communicate and it was discovered that it was a telepathic type. Voices are clearly heard but, their mouths were never opened. One of the abductees said it emphatically that he knows whoever speaks to him among them, whenever a sentence or word is spoken.

One evening, after coming in contact with the news about what the great and retired Israeli space chief, Haim Eshed, said about the existence of the Galactic Federation, I was a bit skeptical about it. I knew very well before this time that ETs exist and I conversed with a few of them but, they did not mention this *federation* thing. So, I was on my laptop in my sitting room, my wife travelled with the kids and I felt this heavy drowsiness that came around 11:57pm. So, I just lay down on the three-sitter couch to take a nap and continue later. I slept for few minutes when I noticed that my two eyes just popped open and I saw these alien guys walked through my wall and right into my sitting room. I wanted to sit-up but, I could not move any part of my body. I quickly counted them, about thirteen in number. They were in different shapes, sizes, heights and forms before me. Honestly, I wanted to get-up and run but, I could not.

That is when I heard distinctly, audibly but, gently, "Take it easy, Doc! We are for peace; don't be alarmed." I heard this from the one in front but, his mouth never opened one bit. Then he gestured with his head towards others standing behind him and said, "Greetings from the Galactic Federation."

Then, it dawned on me that Eshed was right. [There was a little pause]

Then, without opening my mouth, I could hear myself as I said, "You are all welcome! But, you are all different forms and sizes."

The same one in front replied, "Yes! Each one of us represents the different races that make-up the Galactic Federation."

Immediately he finished his sentence above, I uttered, "So, Haim Eshed is correct?"

He replied, "Exactly! We came to clear that doubt you are having and we now take our leave."

I saw them turned round-about to leave but, I realized that they started to vanish from my very eyes. When they finished vanishing I suddenly discovered that I could move my body parts again. This is what shock me most; throughout that conversation, I did not open my mouth, neither did their P.R.O. (I want to believe, since he spoke for them) but, we heard each other very clearly.

CHAPTER III

What Extraterrestrials Can Do

As I was about putting the final full-stop to Chapter 2 above, it was past 11pm during my writing and I heard clearly, "You should tell the world what we can do." I turned, saw nobody but, I could sense something. Then, I started wondering if I'm the only one who can hear from these beings. I did not open my mouth to ask; I just wondered in my mind. I heard this, "Any enlightened human can hear from us. Many of such humans who hear from us are afraid to admit it openly. They just keep mute but, they do hear."

I heard, "We can breed asexually or sexually on our crafts whenever we are on a journey." This instantly brought cloning to my mind and I had this thought to myself, that travelling to another star system might actually become the only avenue where human cloning can be legally acceptable for the sake of scientific advancement. Cloning is an asexual way of breeding and these beings us it on their spaceship. If the science of cloning can be taking place on their spaceship, then, it is not a small craft. That infers that ETs move in mega size spaceships.

Also, this came, "We define a journey by the number of generations that can make it, even before starting it." I told myself that for a cosmic journey to be tagged *well planned*, it is necessary to determine this before embarking on it. For experimental purposes, some scientists tried to look at the minimum number of crew members for a multi-generational space travels to Proxima Centauri b. Arrival at conclusion shows that minimum of 98 people can start and their descendants will reach. That is exactly how these beings mathematically extrapolate numbers needed for the start.

Again, I heard clearly, "Our adaptively big eyes can see kilometers away. Humans can see just about one-third of the length we can see." At this, I told myself, these beings are older on the evolutionary scale than man therefore, it is expected to be so. The voice added, "And we engage our eyes more than man. As man keeps engaging his eyes, his sight will definitely evolve better, with time."

Furthermore, another capacity came, "We can project our minds to go and view a place, before deciding whether to embark or not." The voice continued, "The earth had some good remote viewers before. Some of them are our reincarnates." If I tell you that I was not startled by this particular information, it is a lie. Yes, before going to a place, it pays to know the terrain – so I'm not against going to view it. But, I never expected that some humans we interact with on earth are ET incarnates. I once met a lady that had only stars as tattoos on her body. I told her I like her tattoos and she said, "I did that so that I keep remembering my home." To be frank, I could not process what she said back then; but now, I have.

Next, an interesting point was made, "We can suck a person into our crafts with what is seemingly akin to lacer technology but, not precisely the same working pattern.

However, laser technology is the closest to what is available on earth to describe it." This next sentence is what really makes it interesting. The voice continued, "We can even move a horse, donkey or cattle with this mechanism and drop them off anywhere, easily." At this point, I remembered that some abductees mentioned being lifted by a beam of light, through a hole at the bottom of the spaceship, right unto a sort of examination table. These abductees said they were helpless while this lift was taking place because they could not scream for help. Others said they were in a subconscious kind of mental state during this lift.

Another thing; the voice said, "By means of a special ray, we can immobilize any mammal, including man. This ray is not visible to humans, for it is not in the visible spectrum." It then dawned on me that it is possibly the ray used on me, when the representatives of the Galactic Federation came to authenticate Eshed's claim. If they do not want a man or animal to struggle, this ray will be flashed on that man or animal.

Noteworthy is this one about chip. "Our technological chip implants are glass-like, tiny and very transparent to the ordinary human eye. A scientific art with an intention to make the ordinary human think it is a mere glass," said the voice. At this I remembered a man who claimed he was abducted and on examination, the doctor found a tiny piece of glass in his leg. He was asked if a glass broke around him and he said, "No!" He was operated and the glass was removed. It was then realized that it may not be ordinary glass. The man talked about his experiences to the media; some people thought he was making things up and others accepted his words as the truth. I believed him wholeheartedly then, because I could not read any lie from his facial expressions. Now, I know what was done; a chip to monitor him or something on his body was implanted into him.

Finally, the spaceship manufacturers need to start thinking of this. The voice said, "We have been able to design our ships and crafts to utilize the particulate ions available in the universe. These crafts trap and use these cosmic particles. That is why there is no visible exhaust from our crafts." This is mind-blowing, right? Yes! So, there is no need to start filling the spaceship with any carbonated fuel. This prevents unnecessary heaviness of the spaceship, reduces amount of energy for trust, and allays the fear of ship explosion while in motion. Again, fear of fuel getting burnt-up before getting to destination is not more there and enough room in the spaceship for other things will be attained. There are other advantages of this type of propulsion mechanism.

Honestly, I believe that all these are revealed for the benefit of humanity, to prop man to crack his brain more and to make us understand that interplanetary travel is very much possible.

CHAPTER IV
The Galactic Federation

This is the cooperatively united different races of some ETs into a galactic family. This family is bound by an official agreement that supposedly favors each and every racial member of this family. The federation sees to it that no racial members' territory is oppressed or attacked in any way. Before this union came to be, there was struggle for superiority among ETs and this struggle led to various disasters. Trust me; there was war among the 'gods.' Put perfectly, there was war among the different races of ETs and it wreaked havoc among them. This war annihilated lives on some planets. Could it be the time that the red planet, Mars, lost its inhabitants?

Following the chaos and wreckage, the surviving ETs had to form a body, an alliance, a union of diverse races of them that are willing, so as to protect one another and equally protect the galaxy. Rather than fight one another, it became working harmoniously together to advance galactic scientific knowledge that will help easy interplanetary expedition. The Galactic Federation believes in 'one consciousness.' That is the binding force that has brought many races of ETs together for the sake of advancing *non-terrestrial technical know-how* that would be used in defending every member race, if need be.

I heard clearly, "The Galactic Federation is real but, the earth is yet to be part of it." The reliable source who spoke to me continued, "Humanity is not yet part of it because humans have not understood the concept and business of one consciousness universe." The source finally said, "The Israeli space chief is right and we wish people can take him seriously." At this juncture, it came to me that the entire world population has not accepted the truth about the existence of ETs. So, how can they come to terms with this *one consciousness* doctrine? This is because to truly understand the meaning of *one consciousness,* then, the people must have known fully well that it is the acceptance of other worldly beings and full readiness to work with all the races of these being combined. The earth will first have to accept that there are aliens out there, before joining the Galactic Federation after understanding the essence of joining the galactic family.

On 8th December, 2020, the media and some individuals wrongly dragged one of the most respected space security chiefs in the history of humanity, Haim Eshed, for mentioning the Galactic Federation and how humans are being awaited to gain understanding of space and spaceships. This man is a respected professor and retired general, who has headed Israel's Defense Ministry's Space and therefore, he did not just wake-up on the wrong side of the bed to say what he said. He revealed another thing from his words; humanity does not even understand space and spaceships yet and therefore, is still yet to be qualified to join the concerned federation. [We hope to get there soon.]

The concept of *one consciousness* is being spoken of by the reputable Dr. Steven M. Greer, a man who has interacted with ETs on many occasions, sat in their crafts, conversed with quite a number of them and bringing the knowledge of peaceful contact to lime light. He is doing pretty well in this field. But, I was a bit alarmed when an ET said, "Dr. Greer is doing well to present *one consciousness* to the public. Some are listening but, others think he's blabbing some spiritual nonsense." Since I have this opportunity, I must say that what Dr. Greer is doing is not a spiritual thing. He is simply presenting other advanced physical entities which we think are far away to us. He developed the protocol for peaceful contact and so many individuals around the world are using it to reach out to ETs.

If we cannot handle the truth about the existence of ETs, how then do we embrace the reality of *one consciousness* from different cosmic races of existence? Some years back, during a fundraiser campaign, the applaudable former president and CIA director, George Bush Senior, was asked when the US government would tell Americans the truth about UFOs. "Americans can't handle the truth," was what he uttered, before the organizers halted the questioning. So, because the truth is not handled, it means that the world space technology may not speedily grow, which will translate to our not quickly arriving at the *one consciousness* and that means the earth may not be part of the Galactic Federation anytime soon. However, it is better we do otherwise. We must accept these truths that look at us in the eyes; UFOs are real, UAPs are from ETs and these ETs are far advanced than us in technology. Because our technology cannot make it to their planets yet, does not mean they do not exist. Because we cannot accurately interpret their signals yet, even when they deliberately sent these signals, does not mean they have tried to contact us yet. Because we saw their spaceship designed to look like a meteorite but, not necessarily one and which does not behave like one, does not mean that we must force it classified as a meteorite. Oh earth, oh humans, there are civilizations out there greater than us. They can reach us but, we cannot reach them yet. Let us embrace this truth and work harder, technologically.

Another thing was revealed by an ET guy. He said, "We have hybrids in the human population. They are conceived like normal children but, some modifications take place before their birth. They are people who think like us and do things like us." This ET guy continued, "Names will not be mentioned but, there are many of them. By the virtue of our hybrids, science is advancing and we are pleased that one day, humanity will see us as we are." Look at that; by the virtue of our hybrids science is advancing. This reminds me of two guys I met about 5 years ago. Both of them told something that baffles me till today. They both said that sometimes, they have lucid dreams of watching people construct things. Most of the time, the things constructed have never been made before. On waking-up from the dream, if they gather the materials and construct those things which they dreamt about, they end up inventing something new. Could it be that such guys are the hybrids of these ETs? Maybe!

I was told the advantage of being a member of the Galactic Federation. The voice said, "The good thing about joining the federation is that you become a family and being a family, other members will rally to your side in times of galactic need." This is pretty self-explanatory.

CHAPTER V

Will Extraterrestrials Invade The Earth?

No race of ETs has plans to invade and colonize the earth. So far, the different races making-up the Galactic Federation are patiently waiting for humanity to grow into the 'adult' who would then be initiated into the galactic adulthood i.e. registering with the universal body. Let us look at from this perspective. What would you look at and envy in the hands of a toddler that you would end-up usurping, while you are an adult with everything? These beings are the galactic adults who have everything. Earth is still a technological toddler; we are not yet on the Kardashev scale type 1. So, what will a type 3 civilization want with us? Nothing! If they need anything, there are planets with liquid methane out there, others rain diamonds etc. and they simply mine these things from these places, without stress. They would never come to the earth where they have to dig, waste their time and sweat to get things easily gotten on other planets. These are evolutionary hyper intelligent beings; not dull beings. So, they will never invade the earth. Those trying to form a coalition against ETs invasion, well, it is actually a vain thing because the invasion will never take place.

Another perspective is this – as put earlier in the preface – what concern has lions with ants? Let us assume you are a lion now, will you envy the home of ants that much that you will want to drive them out and take over it? I bet no! These guys existed before us and are thus, further than us in every facet and aspect of life you can think of. They do not envy the earth. As a matter of fact, comparing the earth to some planets out there, it has minimal resources. Some planets are far richer than the earth in terms of mineral resources and are not yet harboring intelligent life. These are the planets being mined and explored by these ETs.

Did you know that these ETs are to nurture humanity until our cosmic awareness is complete? That is right! There are our cosmic elder brothers and sisters. What they are to do is to guide humanity. They do this via earthly human elders that are called the sages. They interact with the sages and tell them the next phase for humanity and how to attain it, if the elders are missing it anywhere. But, as long as the elders are on course, they just watch from afar. Did you know the real reason behind Pythagoras's teaching from behind a curtain? Look deeply! Pythagoras was one of the sages of his time. There are some people who are earth's sages today. These people interact with these beings, give a particular teaching, invent something new etc. but, may not tell you their interaction with these beings.

These beings are in actuality, protecting the earth at the moment. Though this protection is indirectly done but, it is on-going. They are benevolent beings who have attained very high level of enlightenment. As enlightened beings, it is their heart-cry that humanity becomes as enlightened as they are. So occasionally, teachers, earth-minders and earth-cleaners are sent from various races of ETs to various races of men. To buttress this point, there are people in some religious and historical writings whose parents were never known. These people were known to have certain capacities in helping humans and somehow, men and women of their time paid attention to their words to the point that some people paid homage to them. In obeying the teachings and earth-minding advice of these ETs who disguised as humans in the past, the earth is still preserved till date. In reality, they protected the earth, though indirectly.

Try speaking; the earth will not be invaded by ETs. Nobody will leave the city and relocate into the jungle. [That is exactly what the ETs would be doing, if they invade the earth.] Of course, they made paradisiacal changes, consciously to their planets and their ways of life. The galactic adults are rather waiting for humanity to grow and then join them who have been protecting the earth for further strengthening of the star force of the universe.

CHAPTER VI

The Former Extraterrestrials

Every race and virtually every nation has encountered one or more ETs. However, not all were documented and neither did all leave monumental evidences for historical evaluation. Nevertheless, all the visits from ETs were for specific purposes wherever they visited on our planet earth. From legends about these unearthly beings, it was known that some finished what brought them and left without meeting physical death in crafts and witnessed by the humans they mingled with. Some just vanished, without any trace about their where-about after their earthly assignment. Others died like earthly humans and were buried.

The accounts of the visits of other worldly beings are so numerous that if we are to consider so many in this book, then, it will be a digression from the targeted message intended for consideration. A few accounts with historical significance will be looked at here. From all sorts of teachings and enlightening humanity in various fields, to helping man rule over his territory, various ETs with different names have contributed one way or another to this planet, earth.

The Mayans are one of the most notable among the civilizations with well documented chronicles of the visits and activities of their unearthly overseers. The ETs that visited the Mayans came from the Pleiades constellation, mainly. These beings were said to have brought knowledge of Math and Science, according to Mayan archeological history. This civilization did well between 750 – 1,200AD. It was known that the human priests overseeing the Mayans started talking about *going home*. The end of each long calendar cycle of about 400 years each (called Baktun) was celebrated in the days of this civilization. But, about 95% of the Mayans just vanished in the 10th Baktun and this particular Baktun was not celebrated. The surviving Mayans of today are still talking about how their ancestors *went home*; not that they died from anything, neither did they leave because of drought.

The conclusive realization is that at the 10th Baktun, it was time pre-ordained for the *harvest* of the Mayans ready to accompany their unearthly visitors back to the Pleiades star system and 95 out every 100 living Mayans back then, embraced this choice. In the case of the Mayans, their visitors did not return alone; they went back with their host humans. I chose this first to emphasize the fact that only because of effective transportation back to the Pleiades star system, which requires a great deal of speed, will the ETs want humans to accompany them back home. [In the next chapter, the speed of ET crafts will be looked at, in greater detail.]

The Dogon of Mali in Africa, were equally visited by unearthly visitors called the Nommos. These were ETs from Sirius, according to the visitors themselves. The Nommos came to teach all life principles to a people that were then underdeveloped and backward. The Nommos equally taught a great deal of astronomy to their host humans; a fact confirmed by their mention of Sirius as a binary star system even before scientific community could discover it. Today, the Dogon people, after being taught life principles, are advancing in every facets of life. Some primitive circles in the Dogon tribe still seclude their newly appointed king to be visited and consecrated by the Nommos before they assume office.

In China, the yellow emperor reigned throughout the 2,700BC. This emperor introduced acupuncture, bronze coins, written language and other things. This emperor founded the Chinese civilization. According to ancient Chinese historical tales, this emperor was floating through space and saw that the Chinese people ought to be taught life advancement skills and he came to assist them attain them.

 After ruling for 100 years, he did not die but, returned to the stars, riding on a yellow dragon. This emperor in actuality, travelled back to the star system he came from, in a UFO, after finishing what he came to do.

Looking at these few examples, the ETs involved came for specific purposes, fulfilled those purposes and they returned. Yes, they come to nurture and nudge humanity in the direction man should go. After seeing us going the way we should, they do not continue to impose themselves on us; they just leave for their home stars. Putting it differently, the former ETs actually came and schooled man in certain; art, science, technology, skill, way of life etc. and returned to their origin star without forcing themselves continually on the people they came to tutor.

There are some instances of the visits of the ETs that showed and told what is different from the above positive influence they had on groups and races of mankind. It was not a negative influence from them in anyway either. In these other situations, the earthly man helped them rather than the other way round. Yes! An earthly man, especially an enlightened sage, would be met and his opinion on a matter would be sought. That is to say, they are truly beings like you and I; they are not necessarily gods. If they are gods or spirits, they would not come for counsel from a man. However, they visit a man whose level of enlightenment has really appealed to them in such situations. A very good example of this type is when one of the greatest yogis known by humanity, Mahavatar Babaji, was visited in the Himalayan mountain by a race of ETs. This encounter was reported by one of his disciples who witnessed it.

This disciple said that this race of ETs came to seek advice from Babaji on how to handle a particular situation going-on in their home planet. The advice was given and he (the disciple) saw the spaceship returned the way it came.

CHAPTER VII

Additional Extraordinariness of Extraterrestrials

The ETs have very improved eye sight. The improvement in their sight came as evolutionary adaptation from continuous use, over a long time. These beings, in the course of scientific inventions and technological advancement – through the perpetual use of their eyes – have come to have a very keen eye sight. Their eyes are that keen in terms of sight that they can see, with just their naked eyes, the bottom of any water body on the earth from the surface of that water body. While standing on the earth surface, they can equally see through and beyond the clouds/dust of the earth's atmosphere. According to what I gathered, few humans are already developing such sight but, it is still at the infant stage. Humanity will get there, especially among those who engage their eyes perpetually as a generational thing.

The crafts can move with same acceleration and speed in water, as it does in the air. By the virtue of their technological advancement, viscosity of water bodies on earth has no drag effect on their crafts and therefore, even on transition from air to water or vice versa, they maintain same speed or velocity without noticeable change. Can Physics really explain this?

When I was in Chapter 3 of this write-up, I heard during my conversation with one of these guys but, I wanted to leave it till this Chapter. He said, "We now have a craft that can reach Proxima Centauri from the earth in 10 days." Immediately I heard this, my mouth went agape and my eyes widened in disbelief. So, this ET guy said, "I can see you don't believe it." It came out me straightly, "No, I don't!" and he replied, "Let's just leave it there." But that night, I went to bed thinking, what if there is a short-cut from the earth to Proxima Centauri? What if there is even a cosmic highway for this trip? What if there are other holes out there apart from worm-holes for the sake of shortening the travel time? Honestly, the earth has a 'digging' of Physics to do.

In the ancient Hindu texts like the Mahabharata, it is documented that some 'gods' came down in flying houses or palaces. The flying houses here are our point of interest. That was a long time ago and the beings which visited came down from the main spaceship using the smaller flying ones that looked like houses or palaces. If those ETs had flying houses back then, what would be the level of advancement now? Houses which fly (or what seemingly looks like a house in the air) is not an impossible feat. After all, with our present earthly advancement, there is a mobile house called Boxabl. Maybe the next target in the line of housing development would be a flying house. Who knows? So, if back then, a long time ago there were flying houses, how much advancement would have been made by now? As we speak, some ET civilizations have flying cities. I say it again; flying cities.

One of the disciples of Babaji (the great yogi) made mention of an ET race that left some lit touches found by their master in a mountain. This disciple said that their master told them that those lit touches were powered by nuclear sources for over a century and that for another century, they will still be brightly shining. After that, then, decrease in brightness may start being noticed. That is to say that this concerned race of ETs had capacity to harness nuclear energy for over 200 years. That was back then. How much would this same race of ETs do with nuclear energy today? I bet far greater than then.

How about the alien base on the moon found by Ingo Douglas Swann, the American famous remote viewer? In which he saw ETs performing experiments in what was seemingly looking like a laboratory. Though his projected mind went there to see this experimental base but, his mind's presence was noticed by all the ETs who were physically present and busy on the moon. Their sense of detecting the presence of non-bodily being is very much evolved, too.

In 1982, in the deepest lake in the world, Lake Baikal, a research mission was on-going involving seven Russian divers. When these divers went 50 meters deep, they encountered creatures that were unearthly. These creatures were humanoid in form and were appearing to as having a jelly-like helmet on their heads.

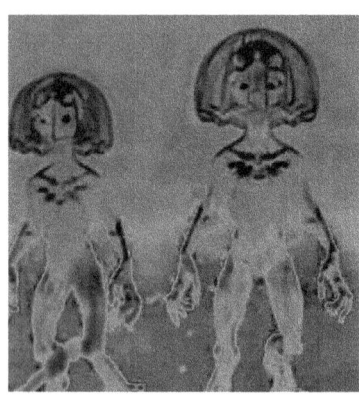

As the divers attempted to capture one of these creatures and bring it back to the surface, there was a sudden force that propelled the divers back to the surface, killing three of the divers. The extraordinary thing noted in this story was that the ETs encountered were not wearing any oxygen mask or tank but, did pretty well at this depth. Also, the force that pushed the humans back to the surface of the water did not give them physical injuries in anyway but, it was that powerful that the 50 meters was covered within a very short time on their way back up that they died from decompression. There is definitely an advanced technical know-how at work here.

Some races of the ETs can shape-shift and that is another extraordinary thing that baffles man. From a known true story, somewhere in Africa, I heard of a being who takes majorly the form of a woman in the night. This 'woman' started being noticed after a crop circle was seen during the rainy season. When this female-looking being started being sighted, the rains stopped completely in that village. It came also to the notice of this village that left-over foods were mysteriously being eaten in the nights and fowls were roasted randomly before dawn at the backs of different houses. This became a subject of concern that a vigilante group was constituted to unravel this mystery. On fateful night for this being, the vigilante chased and pinned her to the ground after she was caught roasting a fowl at 2am behind a house. That was when they realized that there was something off about this woman: her eyes were constantly changing colors and at times, they became as bright as touch-light; she kept changing to male and female in their presence, at a time she changed to a small baby and began to cry; she spoke the local dialect and other unknown tongues and she said that she is not what they think. A physical being that could be held but, these strange things made the vigilante called-upon an old sage available in that village that same night. When he came, he looked at the shape-shifting being and said, "This is one of those from the beyond." He knelt-down and pleaded with the being to leave in peace. This being requested for fire woods to lit and signal her people, who mistakenly left her, to return and pick her. When it was dawn, in the fire woods were seen piled and lit in a strange pattern, the being was seen no more and the rains returned.

Meanwhile ETs are known to pass through earthly walls but, whether it applies to all races or just some of them, is still something to be fathomed. Could it be that these beings from other star systems are made-up of molecules that could easily slip-through our earthly walls? Even the famous Dr. Steven Greer, who is knowledgeable in this field, once said on a TV program that he saw an ET physically on a mountain. This ET was materializing and dematerializing sort of, before his very eyes and then, touched him. A touch he still says he well-felt. These beings can pass through walls and when they touch, those people touched feel it. So, they are not spirits in any way.

Many other relatable and realistic happenings are out there that earthly technological feats would have to demystify in the future. Only concerted efforts will help humanity to get there, if we must.

CHAPTER VIII

Conclusion & Advice from ETs

The earth has recorded a good level of technological feat considering the age of human evolution. The levels of scientific and technological advancement are actually in commensuration with humanity's age but, man is not yet at the peak. "There is more to be done," I'm told. One of the ET guys said, "The great Physicist, Michio Kaku, is right; humanity has just scratched the surface of Physics." This ET sees Prof. Michio Kaku as great and honestly, it is correct. I think the world can start addressing him as *Michio The Great*. [I'm not against that.] Kaku is a theoretical physicist who co-founded the string field theory, an aspect of Physics that opens-up our understanding of the universe.

One thing stands out in all I have gathered so far; that it is time for humanity to advance the next loop of scientific and technological loop. I heard very clearly, "Man has conquered gravity and it is time for humanity to beat time." I am made to understand that this is not just to be done theoretically but, practically. In the light of this, one said,

"Let the world leading scientists, physicists and technologists gather together all those who have ever boasted of warp-drive discovery. If all such people can be gathered together under one platform and if possible, under same working roof, within a decade, the world would be shocked what would be the outcome."

I was made to know that those who are the rate-limiting step to kick-start this project have been born and instead of the scientific community to find fault in any design presented, which will end-up delaying advancement, it is best to gather these people who are presenting diverse designs together and support them. At the end, a working model will definitely emerge.

Another thing I was made to see clearly is this; that the red planet, Mars, is the experimental testing ground of man's capacity to advance interplanetary journey and then, to the stars. It was glaringly clear and it should be now, even to you that humanity has got almost everything it takes to land a manned crew mission to Mars. However, a few of the necessary things are all mentally factored-out already, just awaiting real time production. Instead of delaying, hemming and hawing, it is time to perfect every technology and equally create all mentally designed but, yet to be manufactured technology necessary on Mars. Until man lands on the red planet, he will not be confident enough to advance the practical technology to the stars. So, crewed mission Mars landing is the catalyst that will push man to say, "We can now plan to land on another star system."

Now let us put all these chapters together and see what we ought to see. These beings are benevolent. Yes, they are kind. They want man to quickly grow to their level, as-soon-as-possible. If they truly want to invade the earth, these given pieces of advice will never come. Remember that humanity is not up to their level of scientific and technological equal yet. Also, note that they have been coming and going, right from time immemorial. They did not invade the earth then, when we were just hunter-gatherers and they would not do that now either because they understand what is called *Uniformity of the Cosmic Fabric Matrix*. In this fabric, every star and planet/planets around it play a role in maintaining the consistency of the universal existence. Now let us look at it from this perspective; every sewn thread on the clothing we wear. If there is a break in just one of them, that place becomes the beginning point of ruin of the entire material, if not mended quickly. So, the importance of every star and planet is well understood by these beings. As a matter of fact, they would rather have stars/planets live longer than invade them and shorten their existence. Why? This is because dying stars/planets have catastrophic effects on their immediate surroundings. Now, these beings are concerned about the red planet, Mars, quickly having lives on it again. Lives that will in one way or another, start seeing to the long life of the planet again once again. Did you know that the earth cannot face any of these advanced ETs in battle, if interplanetary war should ever break forth? Why? We ain't gonna do jack shit because they are far ahead. But, let us be calm; there is never going to be an interplanetary war. No alien invasion will take place. No *Mars Attacks!* [Mars Attacks! was a 1996 sci-fi/comedy directed by an intelligent man, Tim Burton, released in the US. In this film, aliens from Mars visit earth and meet the president of the US under peaceful deceptions…]

From my clear judgment point, I do not think ETs will want to invade the earth in anyway. However, you equally be the judge. When I got to this particular part, emphasis was laid on the need for unity in advancing the next technological loop on earth, despite our various languages, backgrounds, races, cultures etc. Nothing should a factor of consideration that would deter this growth. They said and I quote, "The capacity to advance is determined by the strength of the string of unity binding the entire scientific and technological field." That is what I heard. Maybe *Michio the Great* will help all of us to understand that in greater details one day but, you and I know what they mean.

www.ingramcontent.com/pod-product-compliance
Lightning Source LLC
Chambersburg PA
CBHW071114220526
45467CB00004B/1861